SATURN

Also by Elaine Landau

ELAINE LANDAU

SATURN

A FIRST BOOK
FRANKLIN WATTS
NEW YORK/LONDON/TORONTO/SYDNEY/1991

For Bari Braunstein

Cover photograph courtesy of: N.A.S.A.
All photographs courtesy of: N.A.S.A. except: Historical Picture
Service: p. 12

Library of Congress Cataloging-in-Publication Data

Landau, Elaine.
    Saturn / by Elaine Landau.
        p.   cm. — (A First book)
    Includes bibliographical references and index.
    Summary: Using recent findings and photographs, presents
information about Saturn's atmosphere and geographic features.
        ISBN 0-531-20013-2 (lib. bdg.)—ISBN 0-531-15771-7 (pbk.)
        1. Saturn—Juvenile literature. [1. Saturn]   I. Title.
II. Series.
08671.L36   1991
523.4'6—dc20   90-13081   CIP   AC

CONTENTS

SATURN

THE RINGED PLANET

CHAPTER ONE

Saturn is a large and beautiful *planet*. It may be best known for the prominent rings that surround it, which give the planet its unusual appearance. Saturn is one of the nine planets that make up our *solar system.* The solar system consists of the sun and the nine planets that revolve around it, along with numerous smaller bodies such as asteroids, comets, and meteors.

The ancient symbol for Saturn is ♄ . *Astronomers* still use this symbol to represent the planet. Saturn was named after the ancient Roman god of planting and harvest. To honor Saturn, every year the ancient Romans held a joy-filled feast known as the Saturnalia. This festival began near the end of December and lasted for a week.

A view of the ringed planet Saturn
taken by the 120-inch telescope at Lick
Observatory in Mt. Hamilton, California.

During the festival period, prisoners were freed from jail as an act of good will; Roman armies were not permitted to start any new wars; and schools and shops remained closed to enable everyone to enjoy the festivities. Special celebrations were also held. Some historians think that a number of our modern Christmas customs including Christmas dinners, holiday parties, and the practice of exchanging gifts, may be traced to the Saturnalia.

The planet Saturn was first viewed through a telescope by the Italian astronomer Galileo in 1610. When Galileo looked at Saturn through his telescope, he was surprised by what he saw. Instead of being round, the planet seemed to have puffy bulges on both sides. In his description of the planet, Galileo wrote that Saturn had "ears."

As time passed, new and improved telescopes were developed, so astronomers were better able to view the heavens. In 1656, a Dutch astronomer named Christian Huygens also noted the bulges on both sides of Saturn. Huygens realized that what had looked like powder puffs or ears were actually nothing of the sort. Instead, Huygens suggested that the planet was surrounded by a ring. At the time, Huygens believed

Here Galileo and his followers look at
the planets through one of his telescopes.
In the early 1600s, Galileo built a number
of telescopes that sold throughout Europe.

that the ring surrounding Saturn was a solid band of some substance. This was an astonishing discovery. Before Huygens' observation, astronomers had no idea that a planet might be surrounded by a separate ring.

In 1675, a French astronomer named Giovanni Domenico Cassini spent a good deal of time studying Saturn. He observed a dark band in what was thought of as a single ring and realized that the band was really a gap between two separate rings. Cassini thus identified two of Saturn's major rings.

Saturn is the second largest planet in the solar system. Only Jupiter is larger. Saturn's diameter is approximately 74,600 miles (120,000 km). That makes the planet nearly ten times as wide as Earth. From Earth we see Saturn as but a bright "star" in our sky; it takes a telescope to see its rings.

Within the solar system, Saturn is the sixth planet from the sun; Earth is the third. Saturn is about nine and a half times farther away from this blazing hot source of energy than Earth is.

Like the other planets in our solar system, Saturn *orbits* the sun. However, because it is so far away and moves at a somewhat slow pace, Saturn needs considerably more time than Earth to

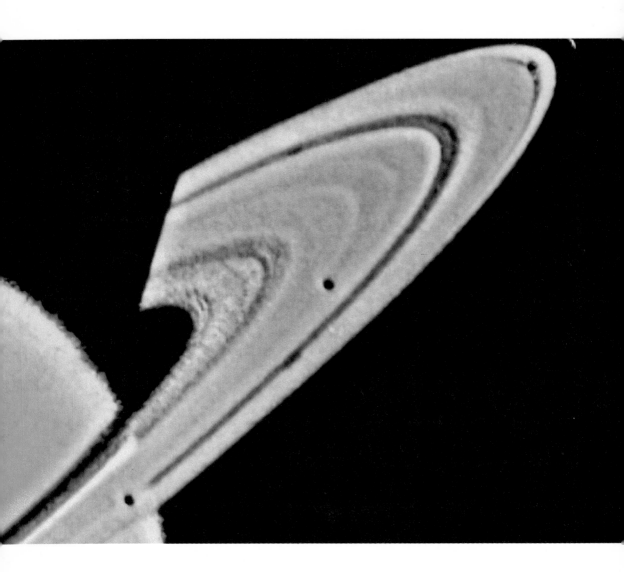

A computer composite of four photos of Saturn.
Different details were used to complete
this modern-day false-color illustration
of Saturn's ring system.

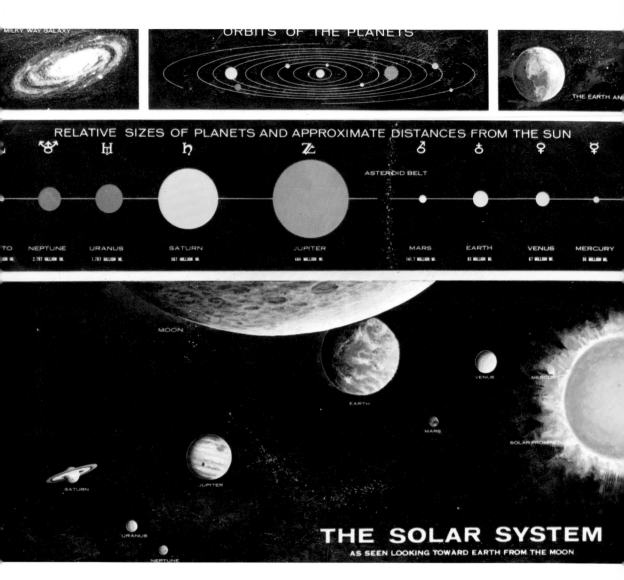

MILKY WAY GALAXY

ORBITS OF THE PLANETS

THE EARTH AN

RELATIVE SIZES OF PLANETS AND APPROXIMATE DISTANCES FROM THE SUN

ASTEROID BELT

PLUTO	NEPTUNE	URANUS	SATURN	JUPITER	MARS	EARTH	VENUS	MERCURY
	2,797 BILLION MI.	1,787 BILLION MI.	887 MILLION MI.	484 MILLION MI.	141.7 MILLION MI.	93 MILLION MI.	67 MILLION MI.	36 MILLION MI.

MOON

EARTH

VENUS

MERCURY

MARS

SOLAR PROMINENCE

SATURN

JUPITER

URANUS

THE SOLAR SYSTEM
AS SEEN LOOKING TOWARD EARTH FROM THE MOON

NEPTUNE

This illustration of the solar system shows the
relative size and orbits of the planets as well as
the distances from each other and to the sun.

complete its journey. The Earth requires 365 days, or one year, to orbit the sun. Saturn needs nearly 29½ Earth years to finish its trip.

At the same time Saturn orbits the sun, it also spins itself around. This process is known as *rotation*. Planets rotate, or turn, on what is called a rotational axis. This is an invisible or imaginary line down the center of a planet on which the planet spins.

A rotation period is the time it takes a planet to complete one full turn. It takes the Earth one day, or approximately twenty-four hours, to rotate once. Saturn spins considerably faster than Earth, completing a rotation every 10 hours and 39 minutes. A day on Saturn would only last a little more than 10 earth hours. The only planet that turns more rapidly than Saturn is Jupiter. Jupiter rotates once every 9 hours and 55 minutes.

SATURN'S ATMOSPHERE AND INTERIOR

CHAPTER TWO

Saturn is actually a large whirling body of gases. It is not a solid planet the way Earth is. You couldn't stand or walk or drive a car on Saturn.

Dark and light belts appear on Saturn's clouds. Beautifully colored bands of pale gold, beige, and white can be seen, although these bands are not as prominent as those of Jupiter. These areas are actually made up of mostly poisonous gases such as ammonia and methane. Some water vapor is present as well. Normally, these gases are invisible. However, because it is extremely cold this far from the sun, the gases have frozen and formed crystals. When we see the dark and light belts on Saturn, we're really looking at the frozen solid crystals of the gases.

Saturn's clouds are not quiet motionless areas. Scientists have learned that a good deal of activity occurs in these regions. There are raging storms and powerful gusting winds. One particularly large storm on the planet may be seen just below Saturn's *equator.* This turbulent storm center appears as a massive rose-colored oval spot on the planet.

The winds on Saturn blow continually at tremendous speeds. In fact, Saturn's winds may reach speeds of over 1,000 miles (1,610 km) per hour. That's more than three times the speed of the strongest winds ever experienced on Earth.

While the Earth's *atmosphere* largely consists of nitrogen and oxygen, the most common elements in Saturn's atmosphere are hydrogen and helium. Some poisonous gases are present in smaller amounts as well. Saturn's atmosphere is also considerably deeper than Earth's. It extends many more miles into the planet's interior.

This photo of Saturn taken from space shows the colorful bands seen on this gaseous planet's moving clouds.

Some scientists believe that toward Saturn's interior the hydrogen and helium on the planet are mostly liquid. The pressure on these elements from the atmosphere above has caused them to remain in a fluid state. The only solid part of Saturn may be found at the planet's innermost center. Here a very tiny core of extremely hot iron and other hardened materials may exist.

Like Earth, Saturn experiences different seasons and temperature changes. Because Saturn takes so much longer than Earth to orbit the sun, Saturn's seasons are much longer. While a season on Earth only lasts for several months, Saturn's seasons are seven and a half years long.

No matter what the season is, Saturn is always much colder than Earth. This is because the giant planet is much farther away from the sun. Saturn is unimaginably cold. At the outer edge of the

This view of Saturn and its ring system was taken from a distance of 8.6 million miles (13.8 million km) from the planet. Storm clouds and small-scale storm spots are present.

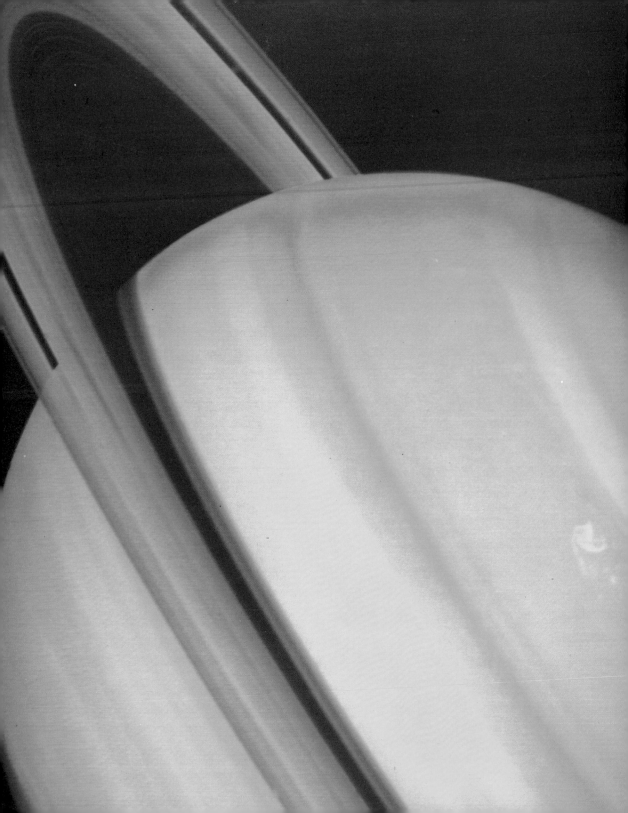

planet's clouds, the temperature may drop to about −285°F (−176°C).

Beneath Saturn's clouds, the temperatures at the inner levels of the planet tend to be somewhat higher. Saturn generates more than twice as much heat as it receives from the sun. Scientists aren't exactly sure why this happens. One theory is that a tremendous amount of heat or energy is worked up as the helium on Saturn slowly sinks through the area of liquid hydrogen within the planet.

Another theory has to do with the planet's creation. Some scientists think that Saturn's internal heat has been present on the planet for billions of years. They believe this source of energy was created when the planet formed. According to their theory, the gases and various other materials that make up Saturn came together with a tremendous impact. This produced a great deal of heat, which remained within the planet, unable to escape easily through the layers of gases that engulfed it.

Over the course of time, some heat has been able to rise to Saturn's outer level. From there, it has escaped from the planet. However, the bulk of the heat, or energy, remains trapped on Saturn. This causes the planet's temperature to be higher than might be expected.

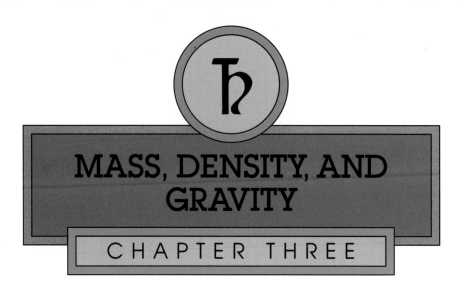

MASS, DENSITY, AND GRAVITY

CHAPTER THREE

Frequently, astronomers will describe a planet in terms of its *mass*. A planet's mass is the total amount of material that makes up the planet. Saturn's mass is 95 times greater than that of Earth. Jupiter is the only planet in the solar system with a greater mass than Saturn's.

Density is another way planets are measured. While a planet's mass is a measure of the amount of material comprising it, the density of a planet tells us how closely packed together these materials are. Solid materials tend to be more tightly packed than liquids and gases.

Earth, which has a solid interior, has the highest density of all the planets in the solar system. Saturn, on the other hand, is primarily a gaseous

This photo shows Saturn's size as
compared with Earth's. Over 300 Earths
could fit inside Saturn.

planet. It has the lowest density of any planet. Saturn's density is only about one eighth as great as Earth's. This means that for its huge size, Saturn is a rather light planet. You might think of Saturn as a giant foam rubber ball. If there were a lake or ocean large enough to set it down on, Saturn would actually float! This is because Saturn's density is less than that of water.

Gravity is an invisible force that pulls everything toward a planet's center. For example, if a circus juggler doesn't catch the objects he tosses into the air, they will fall to the ground. Gravity is the force that pulls them down. The same gravitational force causes autumn leaves to drop to the forest floor as they are shed by trees. Gravity also causes snow to fall on a snowbank.

The same gravitational pull we experience on Earth occurs on Saturn and the other planets in the solar system. The gravitational pull on Saturn is slightly stronger than it is on Earth.

SATURN'S RINGS

CHAPTER FOUR

In 1973, the United States launched an unmanned spacecraft to study the giant gaseous planets of Jupiter and Saturn. The spacecraft, named the *Pioneer-Saturn,* or *Pioneer II,* flew within 13,000 miles (20,900 km) of Saturn in September 1979. The probe sent back valuable scientific data and detailed photographs of the planet and its outer rings.

Two additional space *probes* to study Saturn as well as other planets were launched by the United States in 1977. These were the *Voyager 1* and *Voyager 2* spacecrafts. In November 1980, the *Voyager 1* flew within 78,000 miles (126,000 km) of Saturn. Toward the end of August 1981, the *Voyager 2* flew even closer to this huge planet. It

Here an artist provides a solar
system view of the unmanned *Voyager*
spacecraft's flight as it passes Saturn.

soared within 63,000 miles (101,000 km) of Saturn. The *Voyager* spacecrafts provided scientists with new information about Saturn's rings and moons. They also yielded spectacular photographs of this amazing planet.

Saturn is a stunning sight in space. The planet itself appears as a shining golden globe. But what makes Saturn so magnificent is the series of brilliantly colored rings that surround the planet.

At one time, astronomers using powerful telescopes believed that Saturn was encircled by only three to four rings. Others thought as many as six rings might surround Saturn. However, space probes sent to Saturn proved that they were wrong. We now know that there may be thousands of slender rings circling Saturn's equator.

These rings are closely spaced together. Although Saturn may give the outward appearance of having a few large rings surrounding it, Saturn's major rings actually consist of many smaller parts or ringlets. Some people have compared Saturn's rings to the grooves on a phonograph record.

Saturn's rings are mostly made up of chunks of ice. The rings may also contain particles of dust and rock. These particles vary greatly in size.

This painting shows how the *Voyager*
spacecraft flew behind Saturn's rings
using cameras and radios to measure
how sunlight is affected as it
shines between the ring particles.

A computer-enhanced picture of Saturn shows
the rings and their shadows on a lighted portion
of the planet. The photo was taken by *Voyager 2*
after the spacecraft had passed the planet
and was looking back from a distance of
930,000 miles (1.5 million km).

Some may be less than a quarter of an inch long, while others may be the size of a barn.

Saturn's ring system is extremely wide. The distance across Saturn's outermost ring is nearly 180,000 miles (30,000 km). However, although Saturn's rings tend to be wide, they are extremely thin. Scientists believe them to be less than 10 miles (16 km) thick at most; certain portions are much thinner.

Just as Saturn orbits the sun, the planet's rings orbit Saturn. The rings are made up of a vast number of separate particles, with all the particles traveling in their own orbits. It is as if each particle were a tiny separate moon of Saturn, and together the many billions of these tiny moons form a ring system when seen from a distance.

This computer-produced illustration shows the number and size of the particles within a portion of Saturn's outermost ring. Some particles are as small as marbles while others are as big as a beach ball.

The various orbits of these billions of particles are not all alike. Sometimes particles have collided with one another. These collisions may cause the chunks of ice to break apart. In some instances, the dust covering a particle's surface is thrown off. Following the collisions, splintered pieces may join together to form new particles. Each collision changes the orbit of the bombarded particles. Saturn's gravitational pull keeps the particles from drifting off into space.

Some ring particles may also be kept in place in their rings by tiny moons. These tiny moons, called "shepherd moons," were discovered by the *Voyager* probe.

The *Voyager* probe also discovered spokes, or fingerlike projections, that behave strangely along some of the rings.

In all, the *Voyager* probes provided scientists with a picture of a very complex ring system—and many puzzles to solve.

THE MOONS OF SATURN

CHAPTER FIVE

In addition to its rings, numerous moons orbit Saturn. Astronomers viewing Saturn from Earth with telescopes thought the planet had four major moons. Space probes to the planet revealed that Saturn has at least twenty-three moons. Scientists suspect that there may be others as well.

Many of Saturn's moons or *satellites* are similar. Approximately twenty of them have an outer surface of ice and water. In a number of instances, the moons may largely consist of ice.

Although most of Saturn's moons tend to be round, some are more unusually shaped. A few of the planet's smallest moons, which have only been discovered in recent years, have bumpy oval forms. Some scientists suspect that these moons may actually be the remaining pieces of larger satellites or moons that split apart.

Most of Saturn's moons show numerous *craters* on their surfaces. The craters are actually holes created through collisions with other objects in space.

We'll take a closer look at some of Saturn's more interesting moons.

MIMAS
(diameter: 240 miles; 390 kilometers)

Mimas orbits Saturn from a distance of 115,000 miles (184,000 km) away. Before the Space Age, Mimas was thought to be the second closest moon to Saturn. We now know that there are several small moons between Saturn and Mimas.

One unusual feature of Mimas is the tremendous crater on the moon's surface. The crater is so large that it occupies nearly one quarter of the planet's diameter.

Scientists think that at one time a large object whirling through space collided with Mimas. Mimas might have been bombarded by another satellite or perhaps an *asteroid* (a small planet-like body). If the impact had been greater, Mimas

This painting entitled *View from Mimas* shows
an artist's idea of what it might be like to gaze
at Saturn from its second-closest moon. Notice
how ice has welled up through the dark crater
floor to form Mimas's bright clear surface.

might have split apart. Instead, the collision left the tremendous hole or crater on its surface.

ENCELADUS
(diameter: 310 miles; 500 kilometers)

The moon Enceladus is farther away from Saturn than Mimas. Enceladus orbits Saturn from a distance of 148,000 miles (236,800 km). Unlike Mimas, Enceladus does not have any oversized craters. The moon has some small craters and parts of it seem to have no craters at all. Instead, some parts of the smooth portions have long grooves.

Enceladus looks like a glowing ball in the heavens. Its surface readily reflects the light that falls on it. This accounts for its shining appearance.

TETHYS
(diameter: 650 miles; 1,050 kilometers)

Saturn's moon Tethys is similar to Mimas in some ways. Like Mimas, Tethys's surface is largely covered with craters. Tethys has one especially large crater; this deeply carved-out hole is nearly 250 miles (400 km) wide. There's also an extremely

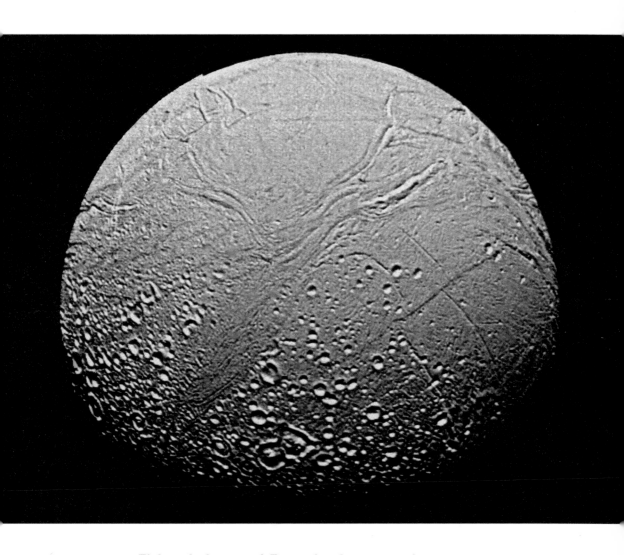

This picture of Enceladus, made up
of several photos taken by *Voyager 2*,
shows the surface details of this moon.

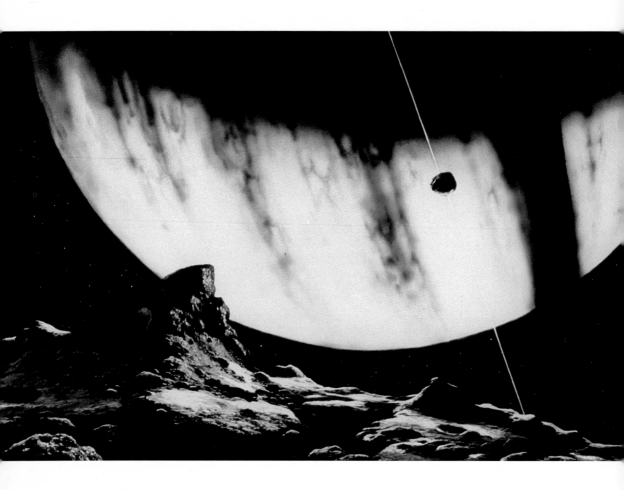

(Above) This painting of Enceladus shows what
it might be like on the frost-covered
surface of Saturn's third-closest moon.
Saturn looms largely in the background.
(Right) A photo of Tethys taken by *Voyager 1*
from a distance of 354,000 miles (566,400 km)
shows the moon's heavily cratered surface.

wide trench or valley on Tethys that stretches around much of the moon. Tethys's surface is largely composed of frozen water.

DIONE
(diameter: 700 miles; 1,120 kilometers)

Dione is only a bit larger than its sister moon, Tethys. Although the two moons are nearly the same size, Dione has a greater density. This means that Dione is heavier than Tethys. For this reason, some astronomers suspect that Dione may contain more solid material or rock than Tethys.

There are some fairly large light areas on Dione's surface. These basins are probably ice. Much of Dione is covered with craters. Other parts have rays or wisps, possibly of crater debris or new ice. Valleys and smooth areas also exist. Astronomers are not sure how these different areas developed.

RHEA
(diameter: 950 miles; 1,530 kilometers)

Rhea is among Saturn's largest moons. Like Dione, Rhea is largely made up of rocky material

This illustration, made up of a number of photos taken by *Voyager 1*, shows Saturn with some of its moons. The artist shows Dione in the forefront with Saturn rising behind it.

Large, bright streaks cross the face of Saturn's moon Dione in this photo taken by *Voyager 1* from a distance of 417,000 miles (661,200 km).

and ice. Parts of Rhea's surface reveal large patches of frost-covered areas.

At one time, Rhea must have collided with numerous objects in space because its surface is thoroughly dotted with craters. In fact, Rhea probably has more craters than any of Saturn's other moons. Rhea's surface tends to be rough in most areas. This is largely due to its plentiful craters, indentations, and ridges.

TITAN
(diameter: 3,190 miles; 5,140 kilometers)

Titan is by far the largest of all Saturn's moons, being the second-largest moon in the solar system. Titan is even bigger than the planets Mercury and Pluto.

Titan is unusual in other ways as well. For example, Titan has its own atmosphere. To date, we know of no other moon that shares this characteristic. Like that of Earth, Titan's atmosphere is largely made up of nitrogen. Its atmosphere also contains ammonia, methane, and some other poisonous gases.

It has been difficult for scientists to learn a great deal about Titan's atmosphere because a heavy fog or haze blankets the large moon. They do suspect that Titan might be made up of a

A painting that shows how Saturn might look
if seen from Titan. Notice the methane
lake in the background as the gas gradually
evaporates into the atmosphere.

rocky core surrounded by a thick icy outer shell forming its surface. Scientists also think that it is extremely cold on Titan. Temperatures may go down to −200°F (−129°C) or even lower. Life as we know it could not exist on Titan. We couldn't survive the cold temperatures, the lack of oxygen, or the poisonous gases.

The three moons that are farthest away from Saturn are called Hyperion, Iapetus, and Phoebe.

HYPERION

Hyperion is irregularly shaped, something like an overgrown squash. Its longest diameter is 250 miles (400 km), its shortest 150 miles (240 km). Some astronomers think that Hyperion may once have been a large moon that was split apart through collisions with other objects in space, resulting in Hyperion's unusual shape.

IAPETUS
(diameter: 890 miles; 1,440 kilometers)

Iapetus is another unusual moon. It is almost two-toned. One portion of Iapetus appears shiny and bright; the other half is considerably darker. But Iapetus is not evenly divided into bright and

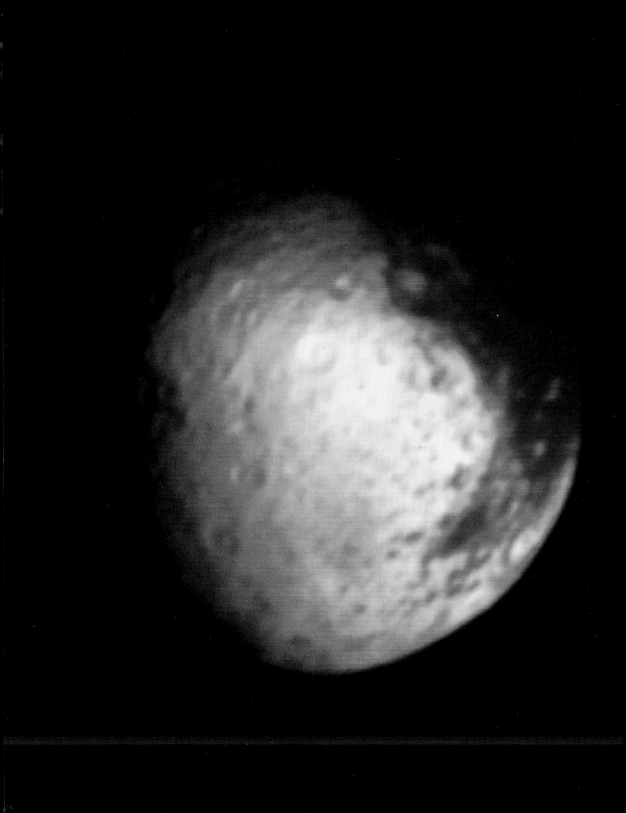

darker halves. Instead, the boundary line between the two regions tends to be uneven and curved.

The light area is white and gleaming, while the dark portion is rusty red in color. Some scientists believe that the bright side of Iapetus may be ice. They think that the darker portions may contain some rocky material.

PHOEBE
(diameter: 120 miles; 200 kilometers)

Phoebe is the farthest known moon from Saturn. It orbits the planet from a distance of eight million miles away. Because of the vast distance between this moon and Saturn, it takes Phoebe over

This illustration of
Saturn's moon Iapetus is
made up of several photos
taken by *Voyager 2*.
Craters may be seen in
both the dark and light
regions of this moon.

a year and a half, or 550 days (Earth time), to complete its orbit around the planet.

Scientists do not have a great deal of information about Phoebe. However, they know that this small moon has an unusually dark surface. Perhaps the most interesting characteristic about Phoebe is that it orbits Saturn in a direction opposite that of Saturn's other moons, or to the west.

Because of its distinct and unusual orbit, some scientists think that perhaps Phoebe wasn't always one of Saturn's moons. Some believe that at one time Phoebe might have been an asteroid. They suggested that perhaps this small asteroid once whirled past Saturn too closely and became trapped by the planet's gravity. From then on, Phoebe orbited Saturn, becoming the planet's most distant moon.

THE FUTURE

CHAPTER SIX

Saturn may be the most dazzling planet in the solar system. In fact, it has often been called the Queen of the Planets. Scientists have learned a great deal about Saturn since 1610, when its unusual nature was first recognized by Galileo. Powerful modern telescopes and space probes have provided us with far greater knowledge about the planet.

Yet astronomers realize that there is still much more to learn. They hope that future space flights will provide the key to unlock the mysteries held by Saturn—the stunning jewel-like planet in the heavens.

A photo taken by *Voyager 2* when the spacecraft was 21 million miles from the planet. Saturn's
moons Rhea and Dione appear here as blue
dots to the south and southeast of Saturn.

FACT SHEET ON SATURN

Symbol for Saturn— ♄

Position—Saturn is the sixth planet from the sun. It is separated from Earth in the solar system by the planets Mars and Jupiter.

Rotation period—10 hours and 39 minutes

Length of year—one year on Saturn is equal to about 29½ years Earth time

Diameter—approximately 74,600 miles (120,000 kilometers)

Distance from the sun (depending on location in orbit)—least: 838,000,000 miles (1,349,000,000 kilometers); greatest: 936,000,000 miles (1,500,000,000 kilometers)

Distance from the Earth (depending on orbit)— least: 762,700,000 miles (1,277,400,000 kilometers); greatest: 1,030,000,000 miles (1,658,000,000 kilometers)

Number of moons—23 known moons. Scientists think that there may be additional moons as well.

GLOSSARY

Asteroid—a small planetlike body

Astronomers—scientists who study the stars, planets, and other bodies in space

Atmosphere—the gaseous outer layer that blankets the Earth as well as some other bodies in space

Axis—an imaginary line through the center of a planet

Crater—a hole or pit on a heavenly body created from an impact with another body

Density—the compactness of materials that make up a planet

Equator—an imaginary circle around a planet midway between its poles, which divides it into two separate parts called hemispheres

Gravity—the force that pulls objects toward the center of a planet

Mass—the amount of matter; the body or bulk of a planet

Orbit—the path in space along which a planet or other body moves

Planet—one of nine large bodies that revolve around the sun

Probe—spacecraft carrying scientific instruments that orbits the sun on its way to one or more planets; in doing so, it may fly past a planet it has been aimed at, orbit the planet, or, in some cases, even land there. Planetary probes collect a great deal of data about a planet even from distances of millions or billions of miles

Rotation—the process by which a planet spins or turns on its axis as it orbits the sun

Satellite—a heavenly body such as a moon that revolves around a planet, or a manufactured device launched from Earth into orbit around a planet

Solar system—the sun and the bodies (planets, satellites, etc.) that revolve around it

FOR FURTHER READING

Asimov, Isaac. *Jupiter: The Spotted Giant*. Milwaukee, Wisconsin: Gareth Stevens, Inc. 1989.

Branley, Franklyn M. *Uranus: The Seventh Planet*. New York: Crowell, 1988.

Daniels, Patricia, ed. *Let's Find out about Outer Space*. Milwaukee, Wisconsin: Raintree Publications, 1981.

Fradin, Dennis. *Moon Flights*. Chicago, Illinois: Children's Press, 1985.

Moore, Patrick. *The Space Shuttle Action Book*. New York: Random House, 1983.

Smith, Howard E. *Daring the Unknown: A History of N.A.S.A.* New York: Harcourt, 1987.

Ride, Sally, and Okie, Susan. *To Space and Back*. New York: Lothrop, 1986.

Wyler, Rose. *Starry Sky*. Englewood Cliffs, New Jersey: Julian Messner, 1989.

INDEX

Saturn (continued)
 gravitational pull on,
  25, 34
 heat generated by, 22
 mass of, 23
 moons of, 29, 34,
  35–50, *37, 39–41, 43,
  44, 46, 48,* 52, 54
 only solid part of, 20
 orbit of, 13–16, 53–54
 position of, 13, 53
 probes to, 27–29, *28,
  30, 31,* 34, 35, *39, 41,
  43, 44, 48,* 51, *52*
 rings of, 9, 11–13, *14,
  21,* 27–34, *30, 31, 33*
 rotation of, 16, 53
 seasons on, 20
 size of, 13, *24,* 53
 storms and winds on,
  19, *21*
 symbol for, 9, 53
 temperature on, 20–22
Saturnalia, 9–11

Seasons, 20
"Shepherd moons," 34
Solar system, 9, 13, *15*
Space probes, 27–29, *28,
 30, 31,* 34, 35, *39, 41, 43,
 44, 48,* 51, *52*
Storms, 19, *21*
Sun, 9, 13, 53–54

Telescopes, *10,* 11–13, *12,*
 29, 51
Temperature, 20–22, 47
Tethys, 38–42, *41*
Titan, 45–47, *46*

View from Mimas, 37
Voyager missions, 27–29,
 28, 30, 31, 34, *39, 41, 43,
 44, 48, 52*

Winds, 19

Year, length of, 16, 53

ABOUT THE AUTHOR

Elaine Landau received her BA degree from New York University in English and Journalism and a master's degree in Library and Information Science from Pratt Institute in New York City.

Ms. Landau has worked as a newspaper reporter, an editor, and a youth services librarian. She has written many books and articles for young people. Ms. Landau lives in Sparta, New Jersey.